AT THE TOP OF THE WORLD

THE GREATEST MOUNTAINS ON EARTH

(and how to climb them)

CONTENTS

ALL ABOUT MOUNTAINS

A mountain is a landform that rises up from its surroundings. Although it is usually at least 330 m (1082 ft) higher than the land around it, there is no standard height. It is defined in the dictionary as 'higher and steeper than a hill'. A hill becomes a mountain simply when somebody names it as one.

This explains why Mount Tenpo in Japan is only 4.5 m high, yet Cavanal Hill in Oklahoma is 727 m high.

Mountains account for around 5% of the planet's land surface. Despite their awe-inspiring stature, they are fragile ecosystems. Many have permanent coverings of snow and glaciers, which provide important water sources. In fact, over half the world's fresh water originates in mountains. The Himalayas have so much ice they are sometimes called 'the Third Pole'. Climate change is causing rapid glacier melt and disruption to the delicate ecosystems of mountains and the surrounding land.

Most mountains form part of a range. Only occasionally is a peak completely isolated. These ranges form natural boundaries that can be hard to cross. Different ecosystems frequently develop on either side of a mountain range, with plants and animals that are very distinct. This is also true for humans – mountains often form geopolitical boundaries that divide the people on either side.

Some mountain terms:

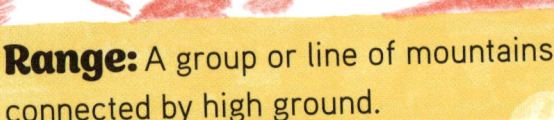

Range: A group or line of mountains connected by high ground.

Summit:
The top of a mountain.

Timberline / Treeline:
The elevation at which tree growth stops. Its location depends on temperature, soil, latitude and other factors.

Shoulder: A flat area on the ridge of a mountain.

Col: Also known as a saddle, the col is the lowest point between two mountain peaks.

Headwall:
A sheer face of rock near or leading to the summit.

Gendarme:
A pinnacle of rock on the ridge of a mountain.

Mountain system / Mountain chain:
A group of mountain ranges. The Rockies and the Alps are mountain chains. The longest mountain chain by far is the Andes.

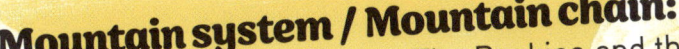

Aiguille:
A French word meaning 'needle', referring to a steep, sharp peak.

How are mountains measured?

Mountains can be measured in three ways.

1. From sea level to peak (the most common way).
2. From base to peak.
3. From the centre of the Earth to the peak.

Mount Everest is the highest mountain the world if measuring from sea level, but Mauna Kea in Hawaii is the tallest if measuring from base to peak, as it extends deep underwater. The peak of Mount Chimborazo is the furthest point from the centre of the Earth.

TYPES OF MOUNTAINS

There are five types of mountains that are formed in different ways, each with a distinctive look.

Fold mountains:

These are the most common type of mountain. The Himalayas, Alps, Andes, Rockies and Urals are all fold mountains. They are formed when two tectonic plates collide head-on and their edges crumple, the way a piece of paper folds when you push against it.

Dome mountains:

Sometimes a volcano doesn't erupt through the crust, but rather leaks lava underneath, pushing up a lump of earth. This type of mountain is called a dome mountain and it usually has a rounded shape. The Black Hills of South Dakota are dome mountains.

Fault-block mountains:

Sometimes, faults in the Earth's crust pull apart from one another, shifting blocks of rock up and down, stacking them on top of one another. These mountains are often very steep on one side and slope down on the other. An example of a fault-block mountain range is the Sierra Nevada in North America.

Volcanic mountains:

As the name suggests, these mountains are formed by volcanoes. When magma deep within the Earth erupts through the Earth's crust, it builds a cone of rock, layer upon layer. Examples of volcanic mountains include Kilimanjaro and Mount Fuji.

Plateau mountains:

Plateaus have steep, mountainous slopes, but are topped with a flat surface, like a table. They can be formed by volcanic or tectonic activity, or by erosion from glaciers or rivers. Plateau mountains are often found near fold mountains. An example is the Altiplano Plateau, which is situated between two Andes ranges.

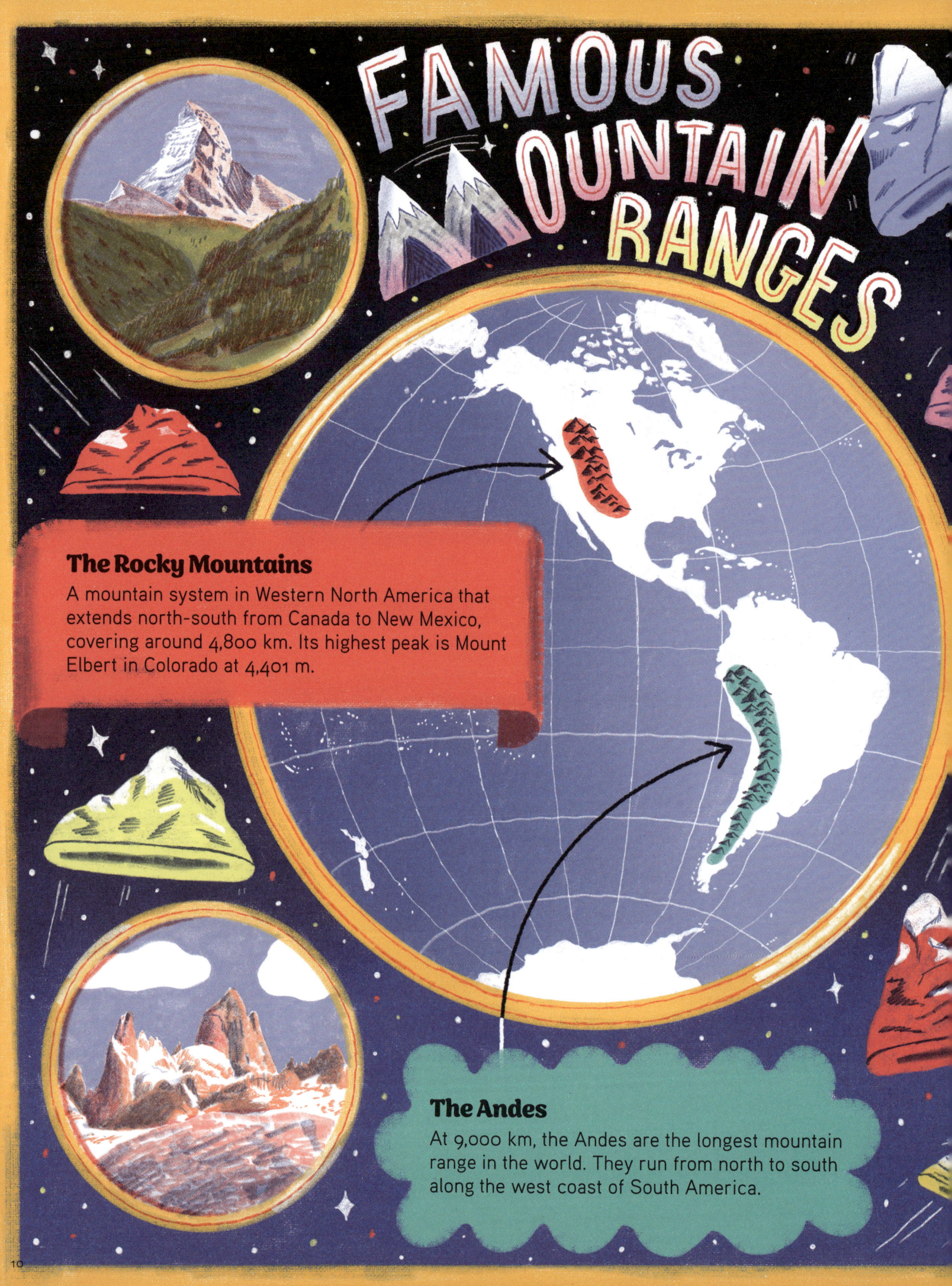

FAMOUS MOUNTAIN RANGES

The Rocky Mountains

A mountain system in Western North America that extends north-south from Canada to New Mexico, covering around 4,800 km. Its highest peak is Mount Elbert in Colorado at 4,401 m.

The Andes

At 9,000 km, the Andes are the longest mountain range in the world. They run from north to south along the west coast of South America.

The Alps

A mountain system in central Europe stretching from Austria in the east, through Italy and Switzerland to France in the west. It has numerous big peaks, the highest of which is Mont Blanc at 4,809 m.

The Atlas Mountains

Fold mountains that span 2,500 km across Morocco, Algeria and Tunisia. The range is home to unique plant and animal species such as the Barbary macaque and the Barbary sheep.

The Himalayas and the Karakoram

Translating as 'land of snow', the Himalayas are the youngest and highest mountain range in the world, with an average elevation of over 6,100 m. The range spreads across Nepal, China, Pakistan, Bhutan and India and contains some of the world's tallest peaks, including Nanga Parbat, Annapurna, Mount Everest and Kanchenjunga.

The Karakoram Range adjoins the Himalayas to the north. It is only 500 km long but contains extremely high peaks including K2, the second highest mountain in the world.

MOUNTAIN EXPLORATION

From ancient civilisations to modern-day adventurers, people have been drawn to the challenge and beauty of the world's high peaks.

In Ancient Greece, Mount Olympus was considered the home of the gods and it was believed that no mortal could ever climb to its summit.

In Ancient China, mountains were considered sacred. Monasteries were built on remote peaks so that the Taoist and Buddhist monks could meditate undisturbed.

In the Age of Exploration, between the 15th and 17th centuries, many European explorers set out to discover new lands and seek treasure. Mountain exploration in this period was mainly focused on finding the fastest trade routes or extracting resources. However, some early climbers were scientists, including botanists and geologists, who were interested in studying the natural history of the landscape.

In the Victorian era, from the mid to late 19th century, mountaineering became truly popular, with intrepid adventurers setting out to conquer the highest peaks in the world. A key figure in the early days of mountaineering was Sir Alfred Wills, an English lawyer and mountaineer who, in 1858, published a guidebook to the mountains of Switzerland, introducing the idea of mountaineering as an exciting hobby.

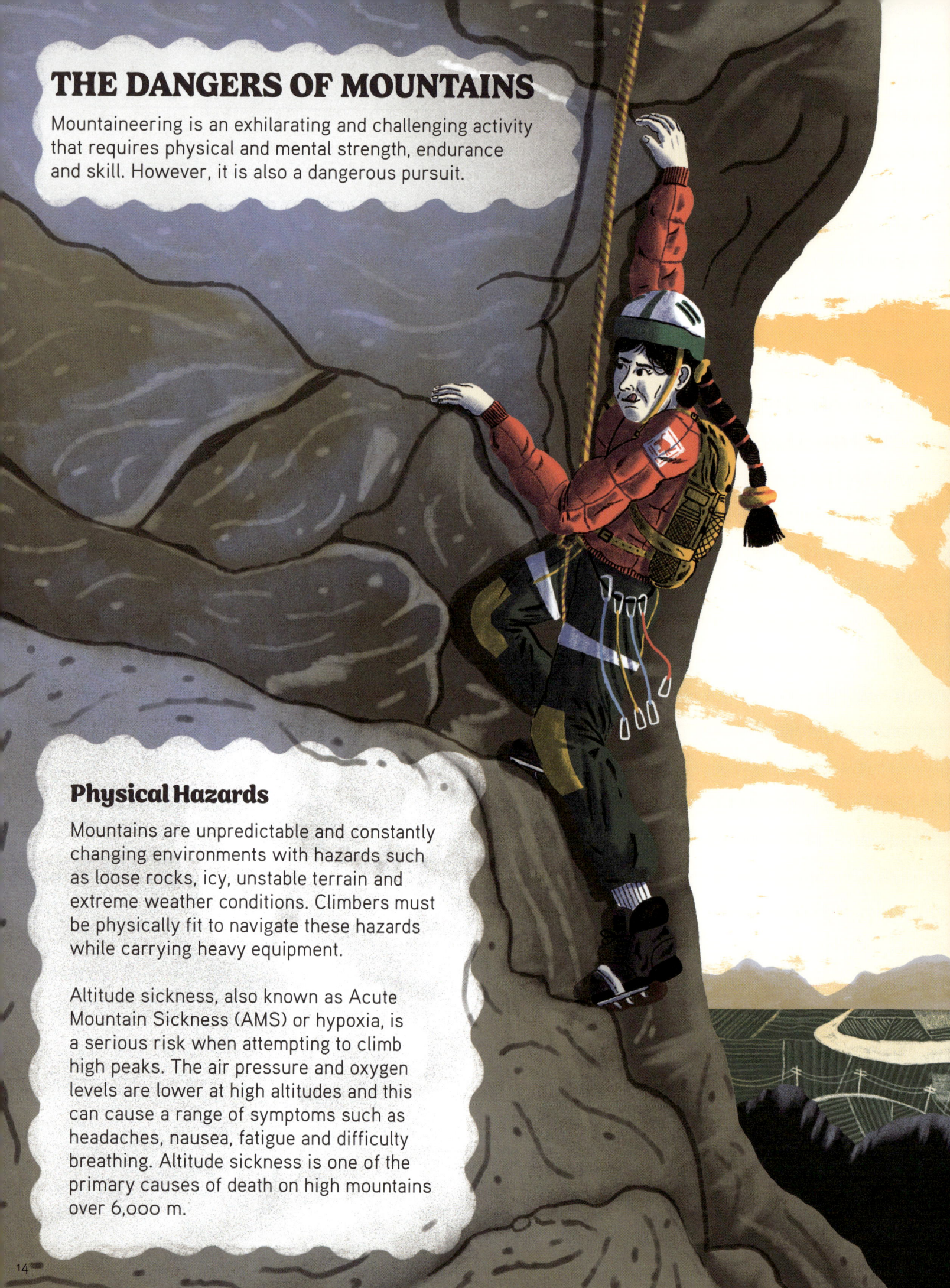

THE DANGERS OF MOUNTAINS

Mountaineering is an exhilarating and challenging activity that requires physical and mental strength, endurance and skill. However, it is also a dangerous pursuit.

Physical Hazards

Mountains are unpredictable and constantly changing environments with hazards such as loose rocks, icy, unstable terrain and extreme weather conditions. Climbers must be physically fit to navigate these hazards while carrying heavy equipment.

Altitude sickness, also known as Acute Mountain Sickness (AMS) or hypoxia, is a serious risk when attempting to climb high peaks. The air pressure and oxygen levels are lower at high altitudes and this can cause a range of symptoms such as headaches, nausea, fatigue and difficulty breathing. Altitude sickness is one of the primary causes of death on high mountains over 6,000 m.

Environmental Hazards

The weather changes quickly in the mountains. Climbers can suddenly find themselves in dangerous conditions.

An avalanche is a huge mass of snow travelling down a mountain slope at great speed. It can be triggered by noise, vibrations or earthquakes. Avalanches can destroy everything in their path (see p.25).

Landslides can be caused by earthquakes or heavy rainfall.

Blizzards of heavy snowfall, strong winds, freezing temperatures and low visibility can present a variety of dangers including disorientation, frostbite and hypothermia.

Earthquakes and eruptions are hazards on volcanic mountains.

Technical Hazards

Climbers often use technical equipment such as ropes, harnesses and other climbing gear (see p.18). Equipment failure can result in injuries and fatalities. Climbers must be skilled in using technical equipment and ensure that their gear is properly maintained.

Human Hazards

Mountaineering also involves a range of human hazards, such as communication breakdowns, disagreements within the team and psychological stress. Climbers must be able to work together effectively and make quick decisions in high-pressure situations.

The presence of other climbers on the mountain can pose a risk. Crowded climbing routes increase the chance of accidents and inexperienced climbers can put themselves and others in danger.

Precautions for Safe Mountaineering

There are some steps that climbers can take to improve the likelihood of a safe and successful climb:

1. Preparedness: Proper training in mountaineering techniques, physical fitness and altitude sickness prevention is vital before embarking on the climb. If travelling without a guide, navigation preparedness is essential.

2. Proper equipment: Climbers should have the appropriate gear for the specific mountain and weather conditions they will encounter.

3. Weather monitoring: Climbers should monitor weather conditions closely and be prepared to adjust their plans or turn back if conditions become dangerous.

4. Communication: Climbers must maintain clear communication with each other and with base camp and have a plan in place for emergency communication.

5. Flexibility: Climbers must assess risks as they arise and if conditions deteriorate they must adjust their plans and never be afraid to turn back. The summit will always be there.

6. Respect for the mountain: Climbers should have respect for the mountain and the environment and take steps to minimise their impact on it.

CLIMBING GEAR
19th Century vs Modern

In the early Victorian days of mountain exploration, climbing equipment was fairly rudimentary. Leather boots, hemp ropes and steel crampons were all that was on offer. Today, waterproof fabrics, lightweight alloys and sophisticated harnesses have made climbing both more comfortable and much safer.

Woollen climbing suit

Felt hat

Canvas and leather rucksack

Ice axe

Snow goggles

Hemp rope

Leather boots

Crampons

Climbing helmet

Nylon rope

Nylon rucksack

Nylon and down
mountaineering suit

Ice axe

Climbing harness

Synthetic
and leather
boots

Crampons

TIED UP IN KNOTS

Tying knots is an essential skill for all climbers, regardless of ability. Your life literally depends on it! There are hundreds of types of knots, but these are a few of the basics.

The Knot

A knot is when a rope is tied to itself. A true knot is capable of holding its form on its own without another object to anchor it.

The Hitch

A type of knot used to attach a rope to an object like a carabiner, post or eye bolt. It is very secure, yet if tied as a slip knot, it can be released quickly with one pull.

The Bend

A bend is is a knot used to join two rope ends.

Practice Rope

Start your practise with a rope between one and two meters long and 6-12 mm thick. This will allow each knot to be clearly examined and also easily untied to allow the rope to be used again. Any flexible rope will do as long as it is not too shiny or slippery.

The Overhand Knot

The overhand knot is the foundation of many other knots. It can be doubled and tripled to increase its strength.

It makes a very secure stopper knot and it can also be tied at the end of a rope to prevent fraying.

A — Loop the rope to make a letter 'e'.

B — Pass the end under and out of the loop. Pull tight to make a regular overhand knot.

C

D

Pass the end through the loop once more and pull tight to create a double overhand knot.

The Figure of 8 Knot

This is another simple and useful stopper knot that is worth remembering. It is easier to untie than the overhand knot.

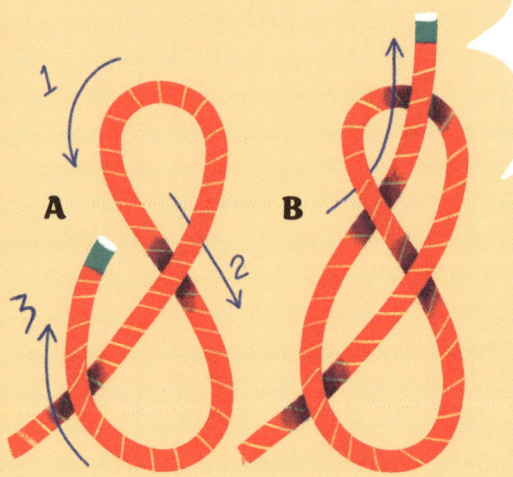

A   B

Make the shape of a number '8' making sure that the rope crosses over itself in the correct order. Reinsert the end back through the top of the 8 and pull tight.

The Clove Hitch

Great for both camping and climbing, the clove hitch can be used to attach a rope to a carabiner, a pole, or even to create a makeshift clothesline when camping.

For a carabiner attachment, start by making two loops, one resting on the other. Cross both loops over and slide the carabiner through the middle. Pull to tighten.

To adjust the position of the knot, loosen it, slide the rope through the carabiner to the new position, and then tighten it again.

A

B

C

The Double Fisherman's Bend

This is two double overhand knots combined, attaching one rope to another. Once pulled tight, it is very difficult to untie.

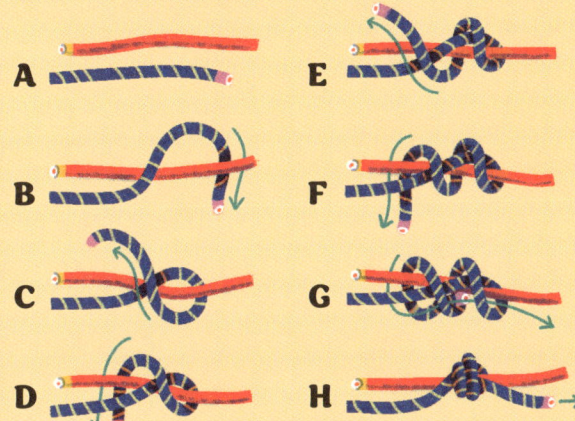

Repeat steps A–H with the other rope to complete the bend.

The Prusik Hitch

This is a slide-and-grip knot. When it's loose it will slide easily, but when pulled, it will grip tightly. This is especially useful as you can use two Prusik knots to climb up or down a rope by loosening one and tightenting the other, inching your way higher or lower.

Tie a loop of rope using a double fisherman's bend.
Then simply wrap the loop around your climbing line four times and pull.

MOUNT EVEREST

Standing at a height of 8,848 m (29,028 ft) above sea level, Mount Everest is the tallest and most famous mountain in the world. It is located in the Mahalangur Himal sub-range of the Himalayas, on the border between Nepal and Tibet.

The Himalayas are the youngest mountain range on the planet. They were formed about 70 million years ago, when the Indo-Australian tectonic plate collided with the Eurasian plate, pushing up a great mountain range of sedimentary and metamorphic rock.

Everest was named after George Everest, a British 19th century surveyor general of India. The Nepalese name for Everest is Sagarmatha, meaning 'Peak of Heaven'. In Tibet, the mountain is called Chomolungma, which means 'Mother of the Universe'.

Physical Features

Everest is a fearsome mountain. Shaped like a three-sided pyramid, its towering peak is so high that the air at the top has only a third of the oxygen present at sea level. The temperature towards the summit fluctuates from -19°c in summer to -60°c in winter and powerful winds of up to 160 kmh batter the mountain year-round. No plant or animal life can inhabit the upper slopes.

The summit is covered with rock-hard snow, on top of which is a layer of softer snow that can range from 1.5 m deep in May to 6 m in September, after the monsoons.

The valleys below Everest are home to various Tibetan-speaking peoples, including the Sherpas (see p.30). The largest Sherpa village is Khumjung, which lies at an altitude of 3,790 m. The Sherpa people have adapted for life at such high altitudes, with genetic factors that allow for a more efficient flow of oxygen through the body.

DANGERS OF EVEREST

Everest is a beast of a mountain; hard to get to and even harder to climb. Many have died trying – even with modern equipment.

Temperatures towards the top are so low that any body part exposed to air is at risk of frostbite. Frostbite is when the skin and other tissues freeze, usually in the fingers, toes, nose, ears and face. In extreme cases, the muscles and tendons beneath the skin freeze, causing permanent damage.

Another major threat is altitude sickness. Elevations above 8,000 m are called the 'death zone', as the oxygen is too thin to support life. It is important to take time acclimatising to the altitude so that your body can increase its production of red blood cells, which carry oxygen throughout the body. If a person were flown to the top of Everest without bottled oxygen, they would die within an hour.

The symptoms of altitude sickness (also known as hypoxia) include headaches, vomiting, dizziness, trouble sleeping and confusion. Bottled oxygen is essential, not just to stave off the physical symptoms, but so that decision-making is not impaired.

The movement of walking, or even just a loud noise can trigger an avalanche of the soft snow that lies on top of the frozen snow layer. An avalanche occurs when a big slab of snow breaks away from a slope and travels quickly downhill, growing and gathering more snow as it cascades down.

What to do in case of an avalanche:

Firstly, try to get off the slab. Head straight downhill as fast as you can and then veer to the side to get out of the way. If you can't get off the slab, reach for a tree and hold on, trying to stay above the cascading snow.

If you can't find a tree, ditch your equipment, roll onto your back with your feet facing downhill, and swim backstroke uphill as hard as you can. The human body is heavier than the avalanche and will sink quickly. You need to stay close to the surface.

Once the avalanche stops, it will settle around you like concrete and it becomes almost impossible to move. As soon as it stops, put one arm across your face to create an air pocket and punch up with the other hand. If you're near the surface you might break through, and if not, it will give you a bigger air pocket.

Once the snow settles, stay still and try to conserve oxygen. 93% of avalanche victims survive if dug out within 15 minutes.

CONQUEST

In the 1890s, two British army officers stationed in India, Sir Francis Younghusband and Charles Bruce, discussed the possibility of an expedition to Everest. They involved two organisations, the Royal Geographical Society and the Alpine Club, but access to the mountain was denied by Tibet.

Irvine

Mallory

In 1920, permission was granted and two attempts were made to climb the mountain in 1921 and 1922. Neither were successful, but one of the mountaineers, George Mallory, identified that the saddle north of Everest was less formidable than it first appeared and this might offer a route to the summit.

In 1924, Mallory made a third attempt. Together with a less experienced climber, Arthur Irvine, he came close to the summit but then disappeared without a trace. His body was only found 75 years later. It was determined that he had died as a result of a bad fall.

Over the next two decades, more attempts were made, with success finally achieved in 1953. A well-organised team established a series of camps up the mountain. Equipped with oxygen and accompanied by Sherpa porters, they climbed the southern route up via a jumble of ice blocks called the Khumbu Icefall. On May 29th, Edmund Hillary of New Zealand and a Sherpa called Tenzing Norgay reached the summit; the first people to have stood on top of the world.

Hillary Norgay

WANG Fuzhou

TABEI JUNKO

Since then, numerous other records have been set. In 1960, a Chinese expedition led by Wang Fuzhou made a successful ascent via the northern route (see p.32). In 1975, a Japanese climber, Junko Tabei, was the first woman to reach the top. In 1983, an American team managed to reach the summit via Everest's trickiest East Face.

TOURISM ON EVEREST

In the last 20 years, climbing Mount Everest has become big business. Around 500 people a day make their way to Everest Base Camp – that's 100,000 people per year! 800 climbers per year ascend the summit. Because there is only a small window between April and May when the conditions are safer, climbers often walk in single file, waiting in line for hours to climb the Hillary Step – the last obstacle before reaching the summit.

With so much tourism, pollution has become a serious problem. 12,000 kg of human excrement is left behind on the mountain each season. Added to this, thousands of kilograms of oxygen tanks, tents and packaging are abandoned by the wayside. The Nepalese government now charges $4,000 per climber for clean-up costs and requires each climber to carry 8 kg of waste as they descend.

However, the bodies of the more than 280 climbers who have died on the upper slopes of Everest have not been removed. They are often victims of falls, so are very hard to reach, and their weight makes carrying them down very difficult. These bodies have become markers on the trail to the summit.

Green Boots is the name given to an unidentified body that is thought to belong to Tsewang Paljor, an Indian climber who died on the mountain in 1996. Sleeping Beauty is the body of Francys Arsentiev, a woman who attempted to climb the mountain without oxygen in 1988. Her body is wrapped in an American flag.

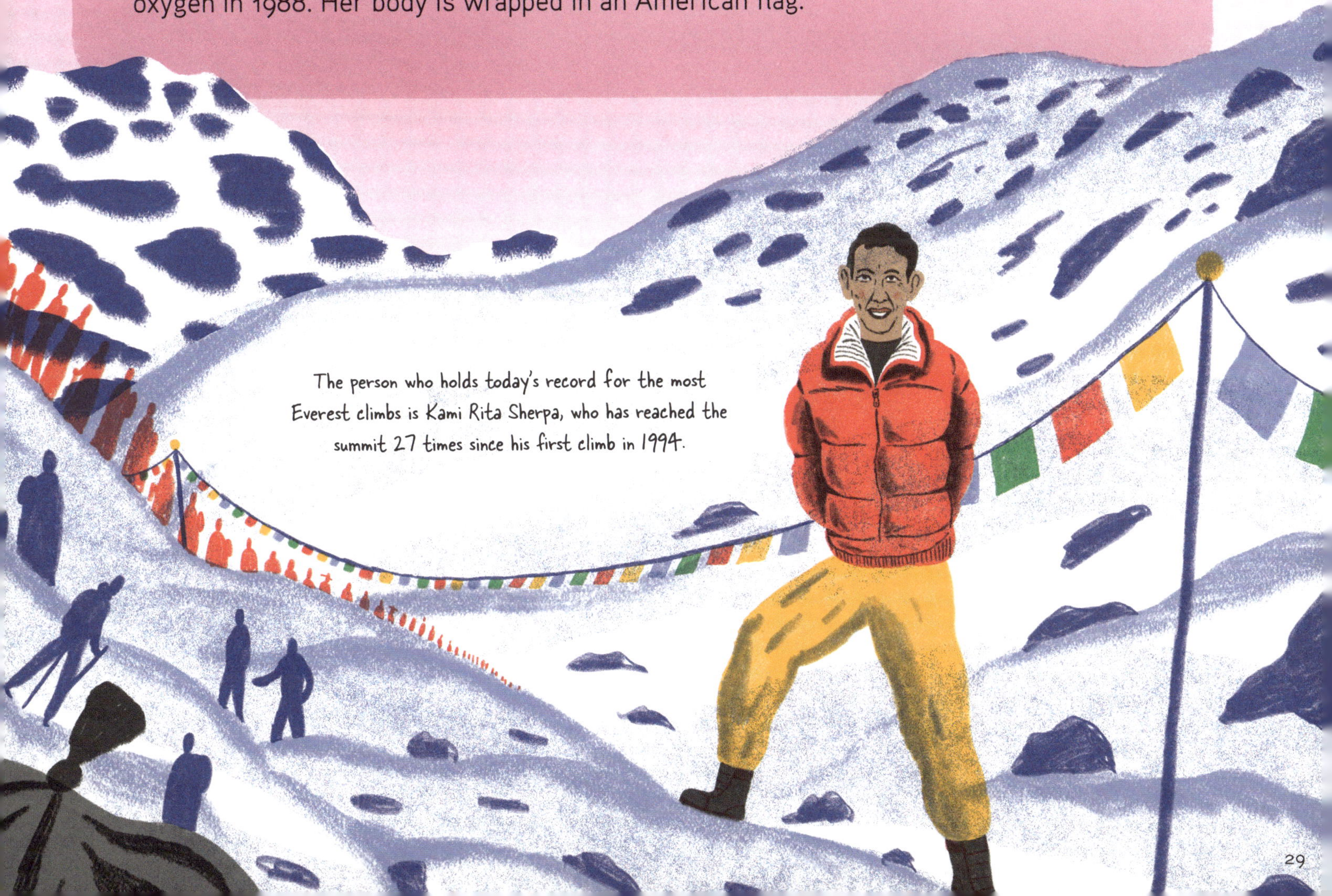

The person who holds today's record for the most Everest climbs is Kami Rita Sherpa, who has reached the summit 27 times since his first climb in 1994.

THE SHERPA PEOPLE

The Sherpas are an ethnic group from the mountains of the eastern Himalayas. Many live in the Khumbu Valley, the gateway to Everest, at elevations of 3,000 to 4,000 m. They are known for their climbing skills and endurance at high altitudes and are an essential part of any Everest climb, acting as guides and porters. They do everything from carrying the loads to setting up camps and securing climbing lines.

Pemba Dorje is the Sherpa who holds the record for speed, ascending to the summit in an astounding 8 hours, 10 minutes.

Sherpas are ethnically Tibetan, and most practise a form of Tibetan Buddhism called Nyingma, or Red Hat. In addition to Buddha, the Sherpa also believe in numerous gods and demons who are believed to inhabit every mountain, cave and forest. Mount Everest is thought to be the home of the great Mother Goddess, Chomolungma. Before each ascent, the Sherpas perform a ritual asking the goddess for permission to climb.

A Sherpa folktale that has found its way into Western culture is that of the yeti. In Sherpa folklore, the yeti is depicted as a shaggy, white, ape-like creature, over 2.5 m tall with a conical head and pointed ears. The story goes that the yetis regularly terrorised local villages, so one day, the village elders hatched a cunning plan.

They brought a great barrel of beer and plenty of weapons to a high alpine meadow, where they pretended to get drunk and fight each other. In the evening, they stumbled home, leaving their weapons behind. The yetis, who had been observing, descended upon the picnic, finishing off the beer and fighting amongst themselves so fiercely that most were killed. The remaining yetis retreated in shame to remote mountain caves, where they still live reclusively today.

THE CLIMB

Despite improvements in equipment and satellite communications, Everest remains one of the most challenging climbs in the world and is only for experienced climbers with a high level of fitness. In Nepal, it is a requirement to travel with a Sherpa.

There are two popular ways to climb the mountain; the South Col Route and the North Col Route:

The North Col Route starts from the north side of the mountain, in Tibet. This route is more technically challenging, with harsher weather conditions and less logistical support. The base camp is higher up with fewer facilities. There is no possibility of conducting rescue operations with a helicopter.

Tibet, China

The South Col Route is the most popular way of ascending Everest. It starts from the south side of the mountain, in Nepal, and follows a gradual ascent to the summit. The trek usually takes about 10 weeks, and much of this time is spent acclimatising to the altitude rather than actual trekking.

2

Camp I is above the icefall and Camp II, known as Advanced Base Camp, is 2.8 km further up.

Base Camp

1

Khumbu Icefall

Camp 1

Camp 2 (Advanced Base Camp)

South Col climbers spend a couple of weeks in Base Camp, which is well equipped with accommodation, food, medical facilities and communication centres. During this time the Sherpas and some climbers will set up ropes and ladders in the treacherous Khumbu Icefall. Shifting blocks of ice makes this one of the most dangerous parts of the climb. Climbers set off at night, when freezing temperatures glue the ice blocks in place.

Khumbu Icefall

5

Just before the summit, climbers must cross the 'Cornice Traverse', a knife-edge ridge of snowy rocks. One wrong step can send a climber down a 2,400 m drop. At the end of this is a 12 m vertical rockface known as the Hillary Step. Only once this has been ascended can climbers say that they have conquered Mount Everest.

The Cornice Traverse

Climbing from Camp III to Camp IV

Summit

Camp 4

4

Camps III and IV are only 1.9 km apart from one another, but trekking at these altitudes is so gruelling, it takes around 12 hours to cover this distance.

Camp 3

3

From Advanced Base Camp, the climbers must use fixed ropes to scale the Lhotse Face. This is the western face of Everest's (almost as terrifying) neighbouring mountain, Lhotse. Climbers use fixed ropes to climb a blue ice wall that rises at a 50° angle, with the occasional 80° bulge.

Scaling the Lhotse Face

The Matterhorn

The Matterhorn is a dramatic pyramid-shaped mountain on the border between Italy and Switzerland. Reaching heights of 4,478 m (14,692 ft), it is one of the highest summits in the Alps.

Known as Mont Cervin in French and Monte Cervino in Italian, the name Matterhorn comes from the German words *Matte* (meadow) and *Horn* (peak). Its distinctive, tooth-like shape was formed by glaciers slicing the mountainside away, creating four faces; north, south, east and west. These faces are so steep that snow slides off them, sending regular avalanches down the lower slopes of the mountain.

The four faces of the Matterhorn

There is a glacier at the base of each face of the mountain. The largest glaciers are the Tiefmattengletscher to the west, part of the Zmutt Glacier, and the Matterhorn Glacier to the north.

Because of its isolated position and great height, the Matterhorn is exposed to unpredictable weather patterns. It often has a plume of cloud off its lee side (the side that is sheltered from the wind). This is known as a 'banner cloud' as it looks like the mountain is carrying a flag or banner.

CONQUEST

The Matterhorn's dramatic beauty has beckoned mountaineers for centuries, but its sheer rockfaces, treacherous terrain and unpredictable weather meant that, in the mid-19th century, it was one of the last great unclimbed peaks in the Alps and had become the subject of an international competition for the summit.

Edward Whymper, a British artist and mountaineer, was transfixed by the Matterhorn. In 1860, he made an unsuccessful attempt to climb it, but turned back just shy of the summit. Over the course of the next five years, he meticulously planned a route, refining his climbing techniques and equipment and making detailed studies of the mountain's weather patterns.

In 1865, Whymper organised another expedition, accompanied by a team of skilled climbers. As he ascended by the Hornli Ridge on the Swiss side, an Italian team made a simultaneous attempt from the Lion Ridge, on the Italian side. On July 14th 1865, Whymper's planning and determination paid off, and his team triumphantly reached the Matterhorn's peak with their rivals only 400 m below them.

However, triumph was soon followed by tragedy. During their descent, one of the climbers slipped and the rope connecting him to the others broke, resulting in four members of the expedition plummeting to their deaths. Only Whymper and two others made it back alive.

THE CLIMB

Although the Matterhorn looks like an isolated peak, it is actually the butt end of a ridge. The most common climbing route to the summit is the Hornli Ridge on the Swiss side, which is not nearly as steep or as difficult to climb as the grand terraced walls of the Italian slope.

Solvay Hut

The base camp of the Hornli Route is called Hörnlihütte and is located at 3,260 m altitude. Around 3,000 climbers summit the mountain each year, and in order to prevent a glut of climbers arriving at once, Swiss authorities have reduced the size of Hörnlihütte so that it can only accommodate 130 climbers per night, with camping outside the hut strictly forbidden. 700 m above Hörnlihütte, the Solvay Hut has just 10 beds and can only be used in case of emergency.

From the Hörnlihütte, it is 1,300 m up the North Face to the summit – around eight to ten hours of climbing. Although the Hornli Route is rated AD (*assez difficile*) – not particularly difficult by mountaineering standards – it is still a challenging climb. The route is exposed to the elements and changeable weather can catch inexperienced climbers unaware. The lower section of the route is lined with loose stones and the upper section is covered with snow and ice, requiring a sure-footed technique. Around five people per year lose their lives on the mountain.

Schmid Route

The imposing North Face route, known as the Schmid Route, is an inviting challenge for more experienced mountaineers. It can only be climbed in specific conditions when sticky snow holds all the loose rock together.

CHIMBORAZO

Mount Chimborazo is an inactive volcano located in Ecuador, in the northern reaches of the Andes mountain chain. At 6,263 m (20,548 ft), it is not nearly as high as Everest, however, it is the furthest point from the Earth's centre. This is because the planet bulges at the equator and Chimborazo is positioned only one degree south.

Chimborazo

Andes Mountains

The Andes run along the western coast of South America, covering about 9,000 km. It is the longest mountain range in the world.

Chimborazo is an inactive stratovolcano. This is a conical volcano that is composed of layers of hardened lava. Stratovolcanoes have low slopes at the bottom that get steeper and steeper as you climb. They are built up over many thousands of years of eruptions and volcanic activity. Chimborazo's last eruption occurred around 550 AD and it is unlikely, though not impossible, that it will erupt again.

Chimborazo has four summits: Whymper, Veintimilla, Politecnica, and Nicolas Martínez. The highest of these is the Whymper Summit. Although it is lower than some other Andean peaks (Aconcagua in Argentina is the highest), it is probably the most challenging climb in the mountain range, due to the glaciers that cover its upper slopes.

CONQUEST

The towering, snow-capped peak of Chimborazo, emerging from its hot, desert-like environs has tempted intrepid explorers since the beginning of the 19th century.

ALEXANDER VON HUMBOLDT

AIME BONPLAND

In 1802, German naturalist Alexander von Humboldt and French botanist Aime Bonpland attempted to reach the summit of Chimborazo. They brought with them around 50 advanced instruments, including barometers and sextants, so that they could measure things like altitude, temperature and air pressure.

Barometers measure air pressure and can help to predict changes in the weather.

A sextant is used to measure angles between the sun, moon or stars and the horizon.

Aided by indigenous guides, they reached an altitude of 5,875 m, where they soon experienced symptoms of vomiting, nosebleeds and dizziness. When they found themselves, weakened, in front of a deep ravine, they had no choice but to turn back. Humboldt was the first to conclude that 'mountain sickness' was related to reduced oxygen at high altitudes.

Although Humboldt never reached the summit, Chimborazo provided the backdrop for his groundbreaking realisation that climate, geography, nature and human societies are all interconnected. His conclusions remain relevant today.

Chimborazo was finally conquered in 1880 by the British climber and conqueror of Matterhorn, Edward Whymper (see p.36), accompanied by two Ecuadorian guides.

Humboldt was a renowned celebrity scientist in his time. To this day, more species of plants and animals have been named after him than anyone else.

THE CLIMB

Chimborazo is a challenging climb that involves snow, black ice and rocky terrain. The risk of avalanches and severe weather conditions is high. Technical climbing equipment and crampons are essential and climbers must begin their journey at night in order to reach the summit before sunrise when the snow melts, increasing the chance of avalanche and rockfall.

El Castillo Refuge

Veintimilla Summit

Ascending the Whymper Summit

The easiest routes up are the El Castillo and the Whymper Routes, which approach from the western ridge. The 12-hour climb is extremely challenging, and even today, less than 50% of climbers make it to the Whymper Summit, with most turning back at the lower summit of Veintimilla. It is such a hazardous climb that ascents without the accompaniment of certified guides are strictly forbidden.

Whymper Refuge

In 2021, an illegal expedition with no guides triggered an avalanche injuring 15 and killing six.

FUJI

Mount Fuji is an iconic symbol of Japan. Located on Honshu Island, about 100 km southwest of Tokyo, this majestic stratovolcano reaches a height of 3,776 m (12,388 ft). It is the highest peak in Japan, dominating the landscape for miles around.

Hokkaido

Honshu Island

Kyushu

Mt Fuji

Shikoku

The mountain forms part of the Fuji Volcanic Zone, a chain of volcanoes that stretches along the southern coast of Japan.

Like many stratovolcanoes, Fuji is made up of several overlapping volcanoes. The oldest volcano is Komitake, which is about 700,000 years old. On top of this is Ko Fuji (Old Fuji) and finally Shin Fuji (New Fuji), which began forming around 10,000 years ago. Shin Fuji is still active. Its last eruption was in 1707, creating Hoei Crater on the southeastern flank of the volcano.

New Fuji ●
Old Fuji ●
Komitake ●

At the summit of Mount Fuji is a crater with a diameter of around 500 m. Around the edge of the crater are eight peaks: Oshaidake, Izudake, Jojudake, Komagatake, Mushimatake, Kengamine, Hukusandake, and Kusushidake. At the foot of the mountain lie five lakes, formed by the damming effects of lava flows. The towns on these lakes offer recreational activities to the millions of local and foreign tourists that visit each year.

At the northwest base of the mountain is Aokigahara Forest, which is said to be haunted by ghosts, spirits and demons.

One of the most captivating aspects of Mount Fuji is its ever-changing appearance throughout the seasons. In winter, a blanket of snow covers its slopes. Spring brings cherry blossoms that cover the foothills. In the summer, the mountain is lush and green and in autumn it boasts vibrant foliage.

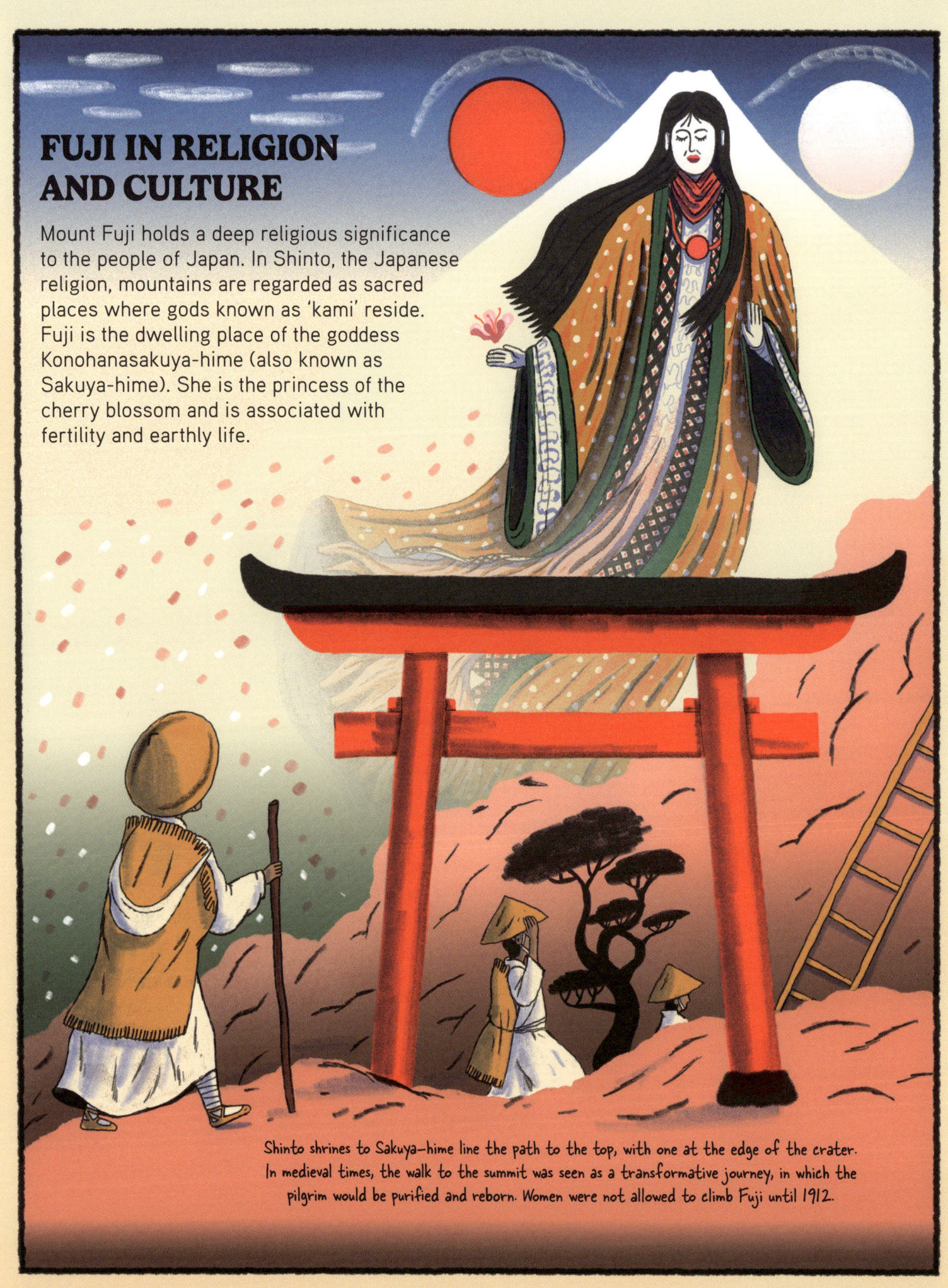

FUJI IN RELIGION AND CULTURE

Mount Fuji holds a deep religious significance to the people of Japan. In Shinto, the Japanese religion, mountains are regarded as sacred places where gods known as 'kami' reside. Fuji is the dwelling place of the goddess Konohanasakuya-hime (also known as Sakuya-hime). She is the princess of the cherry blossom and is associated with fertility and earthly life.

Shinto shrines to Sakuya—hime line the path to the top, with one at the edge of the crater. In medieval times, the walk to the summit was seen as a transformative journey, in which the pilgrim would be purified and reborn. Women were not allowed to climb Fuji until 1912.

Creation Story

There is a famous folk story about the creation of Mount Fuji. A poor, hungry woodsman called Visu lived on the barren plains of Suruga with his wife and family. One night, Visu was awakened by the sound of an incredible crack from deep within the Earth. He rushed to the door to find that Mount Fuji had emerged spontaneously from the soil. Visu prayed and called the mountain Fuji-yama – the eternal mountain. The lands that surrounded it became green and fertile, so that Visu and his family never needed to go hungry again.

THE CLIMB

The trek to the summit of Mount Fuji is challenging but accessible to people of all climbing abilities without a guide. Around 300,000 people climb the mountain each year in the summer months of July to September, when the weather is warmest. Climbing outside of this time is discouraged, as the summit is windy and snowbound in the winter months, with a risk of avalanches.

Yoshida descending route

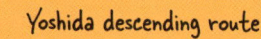

7th Station

Yoshida ascending route

6th Station

The climb is divided into ten stations with the first at the foot of the mountain and the tenth at the summit. Roads go as far as the fifth station, halfway up the mountain, and this is where most people start their trek. The most popular route is the Yoshida Trail, which has large mountain huts to rest in along the way.

The ascent takes five to seven hours. Most people try to time their arrival at the summit with the sunrise (known in Japanese as *goraiko*, 'arrival of light'). To do this, it is recommended to split the climb over two days. The first day is the climb from the fifth to the seventh or eighth station, and the second day sees an early rise and a hike to the summit for arrival around 5 AM.

9th Station

8th Station

5th Station

Some people climb the whole mountain in one go through the night, but this is more dangerous and the risk of altitude sickness is higher if the journey is not broken up.

KILIMANJARO

Mount Kilimanjaro is located in north-eastern Tanzania, about 300 km from the equator. It is a stratovolcano with three volcanic cones. The central cone, Kibo, is the largest. At 5,895 m (19,340 ft), its ice-capped peak is the highest point in Africa. On either side are Shira to the west and Mawensi to the east. Kilimanjaro is the highest free-standing mountain in the world.

Kibo was last active around 170,000 years ago. However, it still has fumaroles in its crater. These are vents that emit gases and vapours but no liquid magma.

Shira and Mawensi are both extinct volcanoes. Kibo is a dormant volcano, which means it has not erupted in a long time but it may erupt again at some point in the future. Uhuru Peak is the highest summit on Kibo's crater rim.

The ice that caps Kibo's summit is the remnants of a much bigger glacier that once covered the entire mountain. In the past hundred years, the glacier has shrunk by 85% due to climate change. There are concerns about the impact this has upon the delicate ecosystems of the mountain, which are closely monitored.

Kibo is connected to Mawensi by an 11 km saddle. Mawensi's cone is jagged and eroded, with a tower-like shape and deep gorges. Below the saddle, Kilimanjaro curves down to the plains below, which lie at around 1,000 m high.

FLORA AND FAUNA

Mount Kilimanjaro has been protected as a national park since 1973 and is home to thousands of animal and plant species. It is made up of four ecological climate zones.

Arctic Summit Zone

At the very top of the mountain, oxygen is thin and ice covers the ground. There is virtually no animal or plant life at this altitude.

Highland Alpine Desert Zone

At higher altitudes, only the hardiest plants survive, such as mosses and lichen, which cover dramatic rock formations.

Moorland Zone

The Moorland Zone is covered by low shrubs, heathers and grasses. Higher up in this zone, the plants have evolved to survive dramatic temperature swings. Senecio are plants that hold onto their dead leaves. As they grow, the old leaves form a fur-like blanket around the trunk.

Forest Zone

The lower slopes of the mountain are encircled by 1,000 km² of dense forest, which provides habitat for thousands of animal species. Monkeys, antelope and bush pigs are common, with occasional sightings of elephants or leopards.

THE PEOPLE OF KILIMANJARO

The Chagga people, who live on the lower slopes of Kilimanjaro, are the third largest ethnic group in Tanzania. They are made up of 400 clans, traditionally led by a chief called a Mangi.

The Chagga farm the fertile soil at the base of the mountain, growing barley, bananas, corn and coffee. Their farming methods of terracing and irrigation have been used for thousands of years. Each Chagga family has a homestead called a *kihamba*, usually in the middle of a banana grove. The house was once a thatched, cone-shaped hut, but now tends to be made of concrete.

Most Chagga today are Christians, however, in the past, their religion was centred around a god named Ruwa, who was believed to live on the summit of Kilimanjaro. The mountain is still considered sacred and old shrines to Ruwa can be seen on the climb.

CONQUEST

Although the mountain's lower reaches have been populated for centuries, interest in climbing the summit of Kilimanjaro only began in the mid-19th century. In 1848, the German missionary Johannes Rebmann reported seeing snow-capped peaks emerging from a tropical landscape, sparking the interest of the Royal Geographical Society, which sent various expeditions to explore the mountain.

Over the following decades, a number of mountaineers attempted and failed to reach the summit. In 1887, a German geology professor, Hans Meyer, reached the lower edge of the ice cap. Lacking the correct equipment for the glacier, he was forced to turn back. Two years later he made a second attempt. This time he set up several campsites with food supplies along the Alpine zone. This meant that if his attempt was unsuccessful, he wouldn't have to descend too far before trying again. On 6th October 1889, Meyer and his partner, Ludwig Purtscheller, made it to the summit of Kibo.

After achieving their aim, Meyer and Purtscheller stayed in the Alpine zone for a further two weeks, observing the landscape and making an attempt to scale the Mawenzi summit. This proved a challenge too far and they were forced to turn back. Mawenzi was eventually conquered by the German climbers Eduard Hans Oehler and Fritz Klute in 1912.

THE CLIMB

Kilimanjaro is one of the only mountains of its size that almost anyone in good health can climb – including kids as young as ten! The mountain is a long, gradual slope. You can simply walk up over several days with no special equipment or previous experience.

There are seven main trekking routes: Machame, Marangu, Rongai, Lemosho, Mweka, the Northern Circuit and Umbwe. There is also the Western Breach – an advanced route for more experienced climbers. All routes start on the Southern side, except for the Northern Circuit. Most routes take between five to nine days to reach the summit.

The easiest route to the summit is Marangu, which is also known as the 'Coca-Cola Route'. With huts to stay in on the way, this is the busiest route to the top. The Lemosho Route, which approaches from the west, is more challenging but also quieter and more scenic.

The main challenge when trekking up Kilimanjaro is the altitude. Uhuru Peak is roughly the same elevation as Mount Everest's Base Camp, and over 50% of those who attempt to climb it suffer from altitude sickness. Altitude sickness is the main cause of death on the mountain.

The second challenge is the weather. At the higher points of the mountain, the sun can beat down during the day, whilst at night the temperature plummets. The best time to climb Kilimanjaro is during the dry seasons; December to early March and June to October. The rainy seasons can be dangerous, especially when hiking on the ice.

DENALI

Denali is the highest peak in North America, rising to a height of 6,194 m (20,321 ft). It is part of the Alaska Range and it is situated in the centre of Denali National Park, a 8,700 km² nature reserve.

Alaska

Canada

Denali National Park

Denali is the tallest mountain in the world from base-to-peak on land. It has two summits. The North Summit is rarely climbed and is slightly lower than the South Summit. Both summits are permanently covered in snow with five glaciers flowing off the slopes of the mountain. The biggest, at 71 km long, is the Kahiltna Glacier. Three weather stations on the Kahiltna Glacier collect data on snowfall, temperature and glacier melt, providing important indicators of climate change.

The mountain's name was only formally recognised in 2015. Until then, it was known as Mount McKinley, a name given to it in 1896 by a gold prospector in honour of the presidential candidate (and later president) William McKinley. In 1975, Alaska changed the name of the mountain to Denali, meaning 'the high one' in the language of the indigenous Koyukon people. It was a further 40 years before the name was recognised by the federal government of the United States.

CONQUEST

From its 'discovery' in the late 19th century, there were numerous attempts to conquer Denali. However, the brutal weather conditions and icy surface meant that for many years no one came close.

In 1909, four Alaska residents: Tom Lloyd, Peter Anderson, Billy Taylor and Charles McGonagall, known as the 'Sourdoughs', set out in the winter months with supplies and dogs.

They took the North Route and discovered a pass, now known as the McGonagall Pass, which allowed them to make it to the North Summit, where they planted a spruce pole.

The group were inexperienced and without photographic evidence, many were sceptical that the summit had truly been reached.

In 1913, the first ascent of the South Summit of Denali was achieved by a group of four climbers: Hudson Stuck, Harry Karstens, Walter Harper and Robert Tatum, accompanied by Esaias George, who was Koyukon and John Fredson, who was Gwich'in.

The Koyukon and Gwich'in are First Nations peoples of Canada and Alaska, who have lived north of the Arctic Circle for thousands of years, traditionally making a living of hunting and trapping.

Fredson was 17 at the time of the expedition and remained at Base Camp for 31 days, hunting caribou and Dall sheep to keep the summit team supplied with food.

During the climb, Stuck spotted the spruce pole on the North Summit through his binoculars. Although the pole has never been seen since, it is widely believed that the Sourdoughs did indeed succeed in their efforts.

Denali Creation Story

The Koyukon people tell the following story about the mountain's origin:

There once lived a warrior called Yahoo, whose great magical power was matched only by his deep loneliness. Searching for a partner, Yahoo set off in a canoe towards the Raven Chief's village, singing songs about his life and his search for love.

Raven Chief hated Yahoo, but many others admired his courage. He caught the attention of one of Raven Chief's daughters. The two fell in love and were married in secret. Afterwards, they swiftly set off in the canoe back to Yahoo's village, hoping to escape the wrath of Raven Chief.

However, Raven Chief caught wind of the news and set off in hot pursuit. Yahoo paddled as fast as he could but Raven Chief used his magic to create a giant storm. Huge waves rose in front of the young couple. As they crested a particularly large one, Raven Chief spotted his chance and threw his mighty spear. Yahoo used his magic to turn the wave to stone. The spear glanced off the stone wave and they were saved.

But Raven Chief did not give up. He paddled around the stone wave and continued to chase Yahoo, who was weak from his efforts. As Yahoo paddled, a wave even greater than the first rose in front of him. Just as he reached the top of the wave, Raven Chief threw another spear. With the very last of his strength, Yahoo once again used his magic to turn the water into stone. The chief's canoe splintered against the great stone wave and Raven Chief turned himself into a raven, flying off defeated.

Yahoo passed out from his efforts. When he awoke, he was back in his village and his bride was smiling at his side. To the west were two great mountains that he had created. The smaller one is called Mount Foraker. The larger one is Denali.

THE CLIMB

Denali is a difficult mountain to climb. It requires fitness, an understanding of fixed line climbing, rope work and glacier travel. There is a high risk of altitude sickness. In addition to this, Denali is one of the coldest mountains in the world, with temperatures reaching -60°C and wind chills as low as -80°C. Avalanches are commonplace.

Denali's remote location means that climbers must carry all their supplies for the entire expedition including food, shelter and equipment. They must plan the ascent carefully and include altitude acclimatisation in their schedules. Despite these challenges, around 1,000 people reach the summit each year.

90% of climbers approach the mountain from the West Buttress Route. This begins at the Kahiltna Glacier and ascends through a series of camps and challenges. The hardest part of the route is the 'Autobahn'. This is the snow and ice slope leading from High Camp (at 5,242 m) to Denali Pass (at 5,547 m). The ground can vary from deep snow (an avalanche danger) to hard ice. More people have died on the Autobahn than on any other part of the mountain. The West Buttress climb usually takes between 17 and 21 days.

Traversing the Autobahn

Tents at High Camp

**Barbara Washburn becomes the first
woman to summit Denali in 1947**

BASE CAMP

PUNCAK JAYA

Puncak Jaya

Australia

The highest peak in Oceania, and the highest island peak in the world, stands at 4,884 m (16,024 ft) above sea level on the island of Papua New Guinea. The mountain's name in Indonesian is Puncak Jaya, meaning 'Glorious Peak'. Its European name is the Carstensz Pyramid, named after Jan Carstensz, the first European to spot the mountain in 1623. His reports of a glacier-topped mountain emerging from a tropical island were much ridiculed, and his sighting was only confirmed two centuries later.

The name of the mountain in the indigenous Amungme people's language is 'Nemangkawi Ninggok' meaning 'Peak of the White Arrow'.

The mountain is located in the Sudirman Range. It has a dramatic shape with great walls of ice dropping 3,000 m below the ridge that connects its northern and southern slopes.

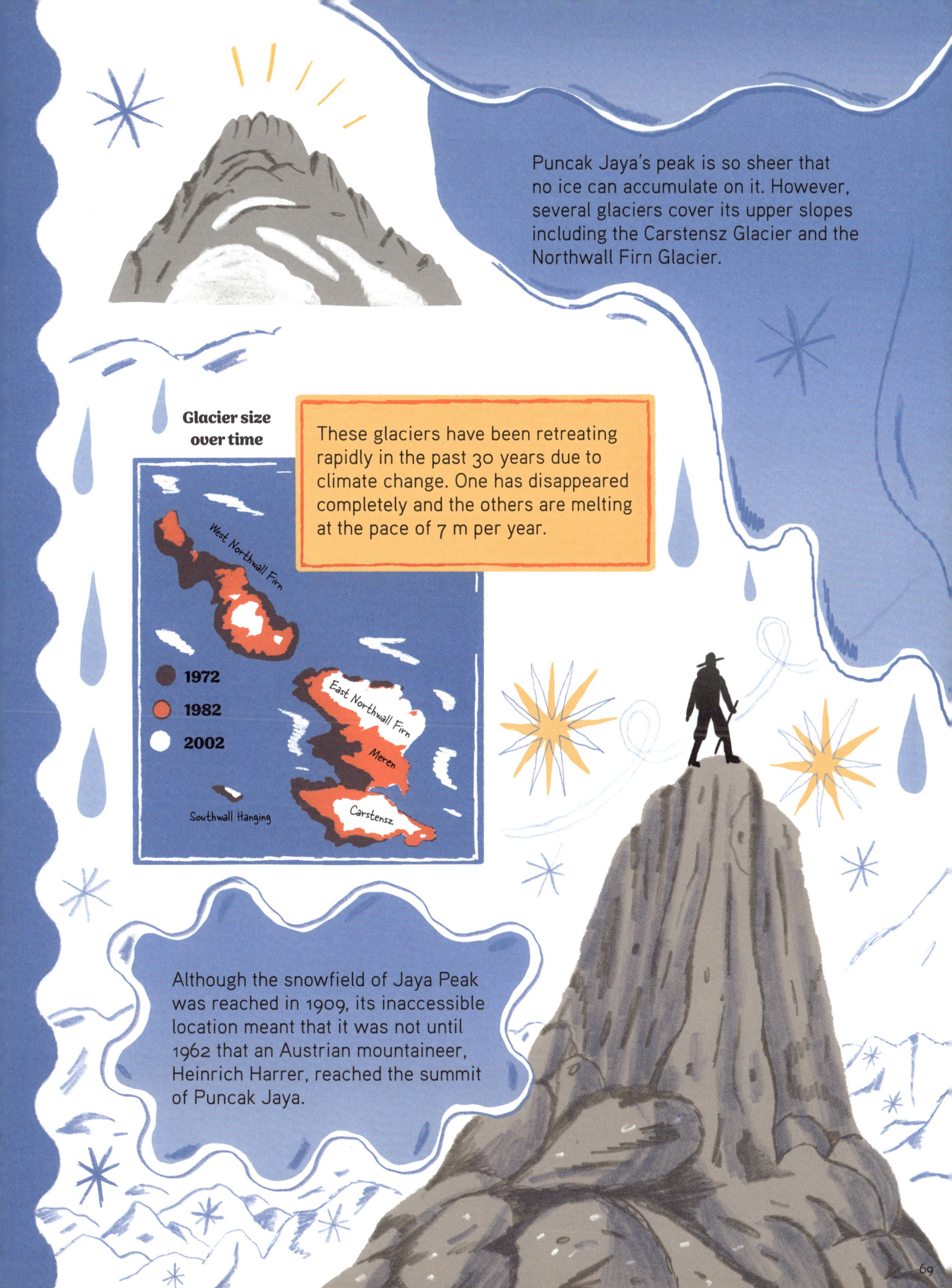

Puncak Jaya's peak is so sheer that no ice can accumulate on it. However, several glaciers cover its upper slopes including the Carstensz Glacier and the Northwall Firn Glacier.

Glacier size over time

West Northwall Firn

- ● 1972
- ● 1982
- ○ 2002

East Northwall Firn

Meren

Southwall Hanging

Carstensz

These glaciers have been retreating rapidly in the past 30 years due to climate change. One has disappeared completely and the others are melting at the pace of 7 m per year.

Although the snowfield of Jaya Peak was reached in 1909, its inaccessible location meant that it was not until 1962 that an Austrian mountaineer, Heinrich Harrer, reached the summit of Puncak Jaya.

CONFLICT

The indigenous people of the area are called the Amungme. Consisting of around 18,000 individuals, the Amungme only made contact with Westerners in the 1960s and many of their ancient customs have been preserved, including traditional dress and a partially nomadic lifestyle.

The term Amungme means 'the first' people.

The Amungme are animistic, meaning they believe that every creature, plant and object has a spirit. There is no separation between gods and nature. The land is therefore sacred and the mountain is revered as particularly holy.

Gold mine

Puncak Jaya

4 km to the west of Puncak Jaya is the Grasberg Mine, the second-largest gold mine in the world. Huge swathes of the landscape have been destroyed in an attempt to extract gold and copper. The mine generates 300,000 tonnes of waste per day, most of which is dumped into the Aikwa River. Native fish have nearly disappeared and the surrounding wildlife and ecosystems have been devastated. This has led to tensions between the Amungme and the Indonesian government.

In addition to these tensions, Papuan freedom fighters and local security forces regularly clash with the Indonesian military, and the situation around the mountain is volatile.

THE CLIMB

Due to Puncak Jaya's remote location and the political instability of the area, not many people have climbed it. It is a challenging climb with the highest technical rating. A helicopter rescue is almost impossible, so only very experienced climbers are advised to attempt the summit.

Most climbers take a small aircraft to a local village on the lower slopes. It is then a five-day hike through dense rainforest to Zebra Wall Base Camp. Tropical conditions mean that it is likely to rain for the entirety of this hike.

From there, a one-day hike leads to Lake Valley Base Camp, where climbers can adjust to the altitude before ascending the summit. The ascent takes 12 to 15 hours and involves fixed ropes along a jagged ridge and a sheer 600 m rock face just below the summit. This is made even more difficult by the near-freezing temperatures at the higher elevations.

The descent from the summit is one of the most challenging in the world. A combination of slippery terrain and exhaustion means that most injuries happen on the way down. It takes three or four days to descend from base camp.

An additional complication can sometimes occur when local climbing porters halt their work to demand (and usually receive) higher pay before agreeing to continue.

VINSON MASSIF

Vinson Massif (or Mount Vinson) is the highest peak in Antarctica, standing at 4,892 m (16,050 ft). It is situated in the Ellsworth Mountain Range, around 1,200 km from the South Pole. The interior of Antarctica has a cold, dry, windy climate, with temperatures that can drop as low as -80°C in the winter months.

VINSON MASSIF

Although there is only around 46 cm of snowfall on Vinson each year, the extremely low temperatures mean that over the years the snow has compacted, forming glaciers that define the landscape of the mountain. The largest is the Branscomb Glacier, which flows down the eastern slopes of the mountain and provides the main climbing route to the summit.

The Union Glacier lies beyond the massif in the Ellsworth Mountains. It is a broad, blue-ice glacier that provides a unique landing strip for aircraft. It also houses the Union Glacier Base Camp, which climbers usually embark from.

CONQUEST

The Vinson Massif was discovered in 1958, but it was not ascended until 1966, when Californian lawyer, Nick Clinch, led an expedition of nine with support from the American Alpine Club and National Geographic. Clinch, who had led the first ascent of Mount Masherbrum in Pakistan, spied a route through a small col onto the upper region of the Branscomb Glacier. After establishing two camps on the way, they successfully reached the summit.

Members of the expedition went on to climb five other peaks in the Ellsworth Range including Mount Tyree (4,852 m), which is a much harder climb.

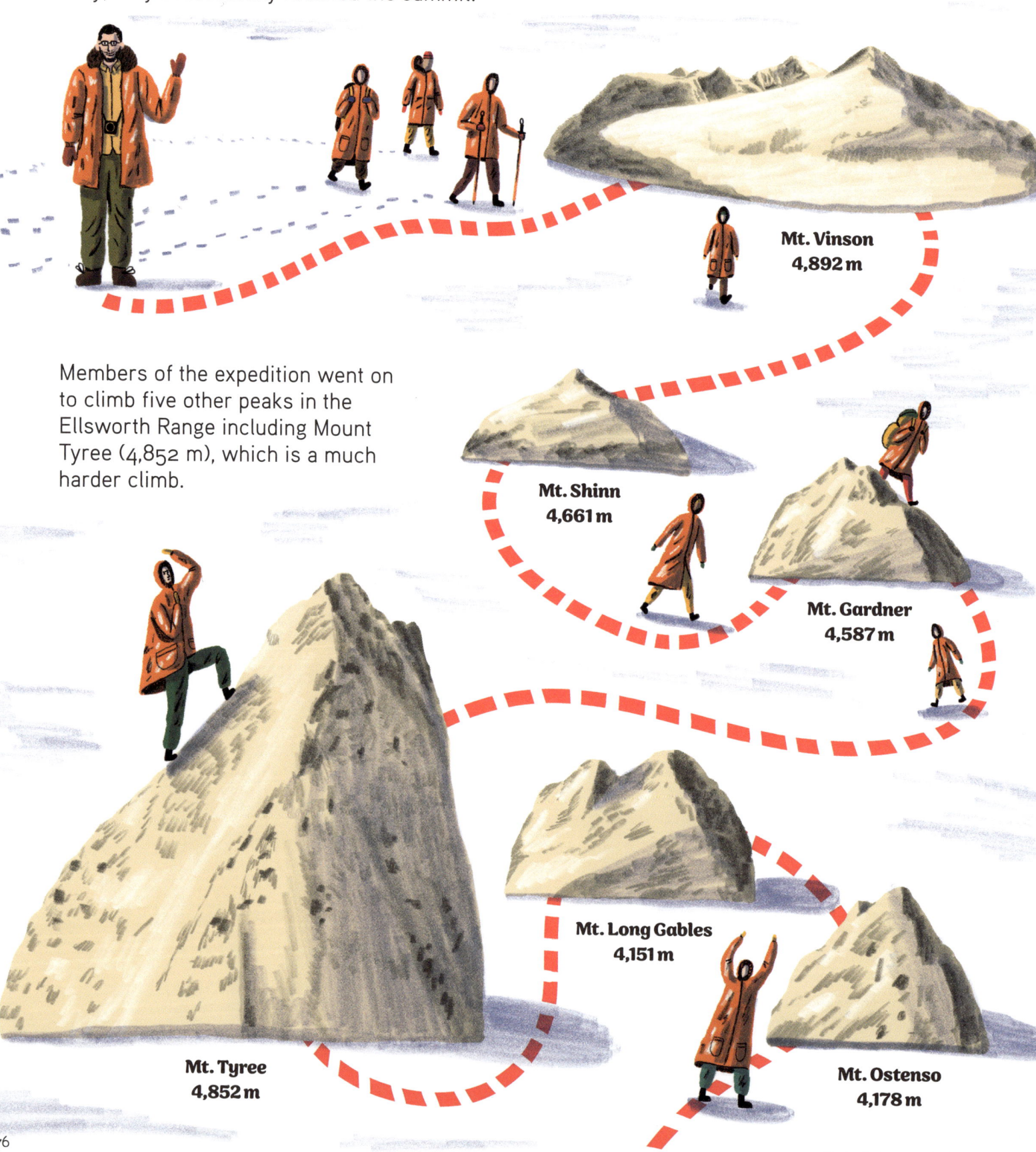

Mt. Vinson
4,892 m

Mt. Shinn
4,661 m

Mt. Gardner
4,587 m

Mt. Long Gables
4,151 m

Mt. Tyree
4,852 m

Mt. Ostenso
4,178 m

THE CLIMB

The only way to get to Vinson Massif is by plane. Most flights leave from the southern point of Chile and fly to Base Camp on the Union Glacier. The climb from there almost always follows the western side of the massif along the Branscomb Glacier. The ascent usually takes five or six days, depending on weather conditions.

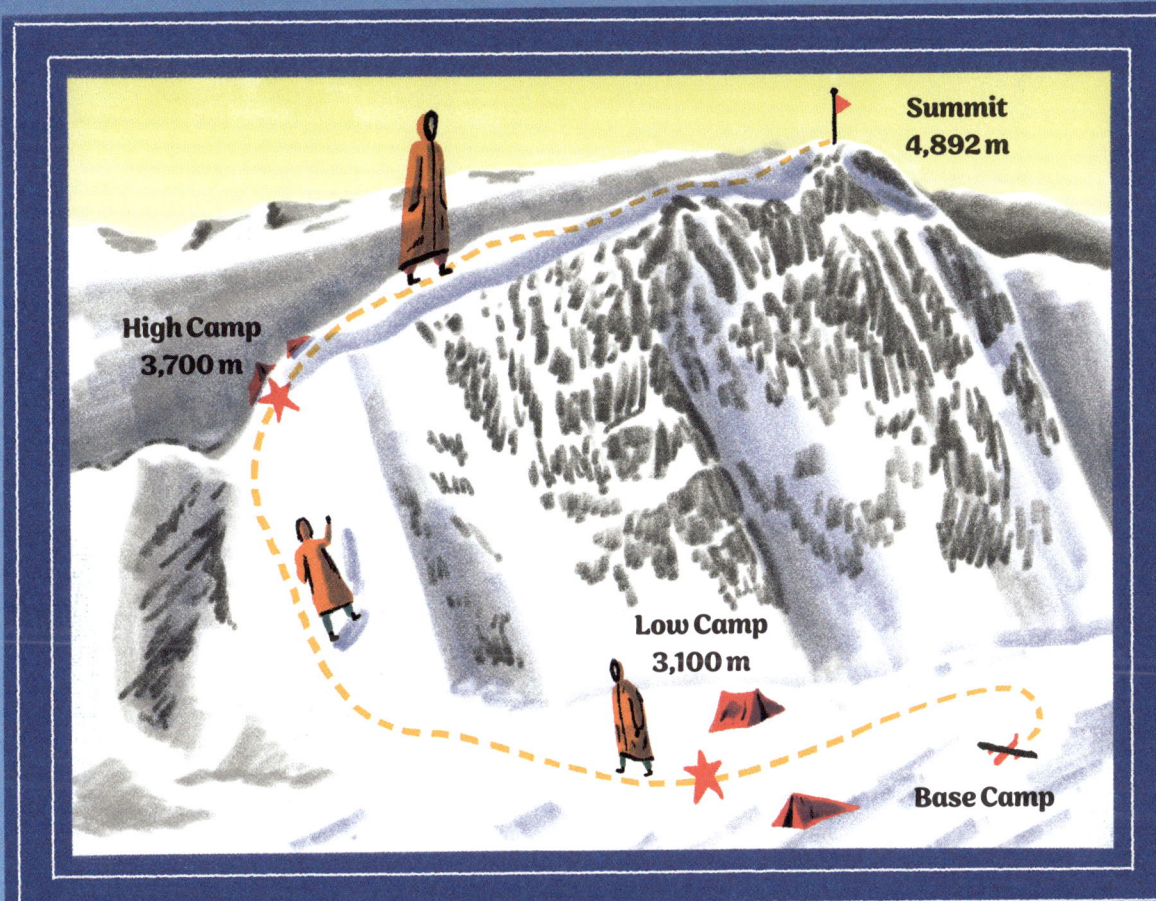

Summit
4,892 m

High Camp
3,700 m

Low Camp
3,100 m

Base Camp

Climbs usually take place between December and February, when there is almost 24 hours of sunlight a day. Temperatures range from -1°c to -24°c, but the wind chill factor can make it feel much colder and frostbite is common. Altitude sickness may also be experienced, but overall, the climb is not too technically challenging. To date, about 1,200 climbers have reached the summit with zero casualties.

GLOSSARY

Alpine: Upland slopes above the treeline.

Alpine style: To be lightly equipped with no bottled oxygen, no overnight equipment and limited food supplies.

Altitude Sickness (AMS, hypoxia): Symptoms of low blood oxygen due to high altitude, including headache, vomiting, disturbed sleep and confusion.

Anchor: A point where the rope is secured to the snow, ice or rock to provide protection against a fall.

Approach: The nontechnical section of the climb that leads to the technical part of the climb.

Avalanche: A rapid flow of snow down a slope.

Belay: A safety technique in which a stationary climber provides protection by means of ropes, anchors and braking techniques to an ascending or descending partner.

Bivouac (bivi): A high camp, not always a planned overnight stop.

Cairn: A pile of rock or wood used to mark a route.

Cam: A spring-loaded device that can be inserted in rock cracks through which a climbing rope can be threaded.

Carabiner: Aluminium devices of various shapes with a spring-loaded gate through which a climbing rope can be threaded.

Clipping in: Using a carabiner to connect to belays and anchors or to connect ropes to protection.

Cornice: Wind-sculpted snow overhanging a ridge.

Crampons: Spiked metal devices that attach firmly to climbing boots to provide reliable footing on ice and firm snow slopes.

Crevasse: A crack in a glacier surface.

Double-wall tent: A tent that protects from condensation by adding an interior wall between the rain fly and the sleeping bag.

Dry-tool: To ascend a section of rock using ice tools, a technique used for short sections of rock between sections of snow or ice on alpine climbs.

Edging: A rock climbing technique where the edges of the climbing shoes are used to stand on small footholds.

Fall line: The direction a fall will take. Avoid climbing in the fall line of another climber higher up.

Figure-eight knot: The basic climber's knot, used to attach a climber's harness to the rope.

Fixed line: A rope anchored to a route by the lead climber and left in place for others who follow.

Gendarme: A mass of rock protruding from a ridge.

Glissade: Descending moderate snow slopes by sliding on one's feet or bottom. Glissading does not work on ice.

Harness: A strong belt with leg loops made of nylon webbing used to secure the climber to the rope.

Hypothermia: Low body temperature caused by the cold.

Ice axe: A mountaineering tool for snow and ice climbing, pointed at the base of the shaft and with a head consisting of a pick and an adze.

Lead: To be the first climber up a pitch and to place protection along the way while being belayed by a partner from below.

Moraine: A random accumulation of boulders, rocks, scree and sand carried down the mountain and deposited by a glacier.

Piton: A metal spike that can be hammered into rock cracks for protection.

Rappel (abseil): To descend a rock face by using a doubled rope coiled around the body and fixed at a higher point.

Rime: A thin layer of ice and hard snow over rock.

Royal Geographic Society: A British institution founded in 1830 to promote the advancement of geographical science.

Saddle (col): The lowest point of elevation between two peaks.

Scree: Small, loose rocks.

Spindrift: Loose, powdery snow incapable of holding protection.

Spur: An outward projecting part of a mountain that curves away laterally.

Timberline: An elevation so high that trees no longer grow.

Webbing: Flat nylon tape or tubing used for slings and harnesses.

Zero: To camp in the same place for a day and achieve no distance in order to rest, acclimatise or to hang out and admire the view.

At the Top of the World

Text by Robin Jacobs
Illustrations by Ed J Brown

British Library Cataloguing-in-Publication Data.

A CIP record for this book is available from the British Library
ISBN: 978-1-80066-047-2

First published in 2024

Cicada Books Ltd
Unit 9, Cliff Road Studios
5 Cliff Road
London, NW5 1UE
www.cicadabooks.co.uk

© Cicada Books Ltd

Printed in Poland on FSC® certified paper

MIX
Paper | Supporting
responsible forestry
FSC
www.fsc.org
FSC® C163799